ZSTU FASHION COLLEGE GRADUATION WORKS 2015

2015 浙江理工大学
服装学院毕业设计作品集

主编 / 陶宁　Chief Editor / Tao Ning　　副主编 / 陈敬玉　Editor / Chen Jingyu

中国纺织出版社

内 容 提 要

　　本书辑录了浙江理工大学服装学院2015届服装艺术设计专业优秀毕业作品，介绍了我院系统完善、独具特色的服装教学体系，记录了学生们如何将创意落在画纸、裁于面料、成为服装的过程，从市场到课堂，从面料到服装，从创意到产品，一步步走来，年轻的步伐轻快而稳健。相信假以时日，这些新生代设计师们的未来指日可待。

图书在版编目(CIP)数据

2015 浙江理工大学服装学院毕业设计作品集 ／ 陶宁
主编 . -- 北京：中国纺织出版社，2015.12
　ISBN 978-7-5180-1841-3

Ⅰ . ①2… 　Ⅱ . ①陶… 　Ⅲ . ①服装设计—作品集—中国— 现代 Ⅳ. ①TS941.2

中国版本图书馆 CIP 数据核字(2015)第 162889 号

责任编辑：张 程 　责任校对：王花妮 　责任印制：王艳丽

中国纺织出版社出版发行
地址：北京市朝阳区百子湾东里 A407 号楼 　邮政编码：100124
销售电话：010 - 67004422 　传真：010 - 87155801
http：//www.c-textilep.com
E-mail: faxing@c-textilep.com
中国纺织出版社天猫旗舰店
官方微博 http://weibo.com/2119887771
北京利丰雅高长城印刷有限公司印刷 　各地新华书店经销
2015 年 12 月第 1 版第 1 次印刷
开本 :787×1092 　1/8 印张 :12
字数 :155 千字 　定价 :168.00 元

ZSTU
FASHION COLLEGE
GRADUATION WORKS 2015

2015 浙江理工大学
服装学院毕业设计作品集

PREFACE
卷首语

序

看到服装学院 2015 届毕业作品结集出版，作为教育者，倍感欣慰。

"春发其华，秋收其实"，服装学院的毕业生们，通过四年的磨练耕耘，积聚出了今天的点点滴滴。

作品虽显稚气，却有年轻人真诚表达的勇气，奋发进取的朝气。

服装，是人的第二层皮肤。

通过服装，我们保护，遮掩，表现乃至定义自己。

服装产业，中国制造的骄傲。

在迈向中国创造的道路上，服装产业的转型升级是极其重要的一环。

服装学院，作为我校最具办学特色和社会影响的学院，其专业设置涵盖了服装产业链的上下游，包括面料、家纺产品设计、服装工程技术、服装设计、服饰品设计、人物造型设计、服装营销、展示与陈列等。

作为国内较早开设服装专业的院校，浙江理工大学的服装学院，为中国服装产业培育了第一代设计师工程师，涌现了 9 位全国十佳设计师，连续十多年获得全国服装教育"育人奖"，是当之无愧的"中国著名时装学府"。

站在如此高的起点，背后有那么多前辈校友的杰出成就，今天的服装学院学子们，将交出一份怎样的答卷？

在这本作品集里，我看到了更大的惊喜。

我看到了学生们力求突破思想观念、手法手段、表达表现的规律模式，用创新的形式语言，畅意创造；

我看到了年轻不世故却源于自我的生活体验，阐释美好的感知和体悟。

这些，都显示了年轻一代的青春锐气与专业才华；

同时，也凝聚了老师的无私喂哺与辛勤汗水。

千里之行，始于足下。

希望这本作品集为浙江理工大学服装学院师生们提供展示宣传的媒介，同时也诚挚期待各界人士批评指正。

浙江理工大学副校长

PART-1

✦ 一 路 走 来 ✦

从百年学府到设计师摇篮

A Hundred Years University to
Designer Cradle

PART-2

✦ 一件服装的生长 ✦

从 种 子 到 结 果

A Process of Fashion Works
From Idea to Creation

PART-3

✦ 一个创意的实现 ✦

我 和 我 的 作 品

Graduation Work
Collection

PART-1

❖ 一 路 走 来 ❖

从百年学府到设计师摇篮

A Hundred Years University to
—— Designer Cradle ——

一百多年前
杭州知府林启的大笔一挥　蚕学馆
浙江引入西式职业教育的第一块基石
植桑养蚕　绫罗绸缎
古老中国献给人类文明的瑰宝
从丝绸到服装　从农耕文明到创意产业
百年学府见证发展　设计师摇篮盛名在外
这里走出了近代中国民族产业的脊梁
走出了许多中国十佳设计师　国际化的品牌
走出了中国创造的新一代力量。

Lin Qi, the magistrate of Hangzhou one hundred years ago,
established the Silkworm School.
It was the first modern vocational school in Zhejiang.
Silk once was the treasure that ancient Chinese civilizations possessed.
Now it is one part of the fashion industry.
The school witnessed the evolution and turned into the designer cradle.
Many entrepreneurs of modern national industry,
the best of China's fashion designers, and brands grew up here.
Now the new generation of China Creation will start from here.

杭 州
自古为丝绸之府

———

蚕学馆
历经百年积淀

———

浙江理工大学
当代的高等时装学府

———

新生代中国设计师
从这里出发

1897
林启 蚕学馆
Lin Qi Silkworm School

∨

1979
浙江丝绸工学院
丝绸美术与品种设计专业
Zhejiang Silk Industrial College
Silk Arts and Variety Design

∨

1982
服装设计专业
Fashion Design

∨

2000
美国纽约州立大学时装技术学院(FIT)合作办学
中美合作项目
Fashion Institute of Technology (FIT)
Cooperative Education

∨

2004
浙江理工大学服装学院
School of Fashion Design and Engineering
Zhejiang Sci-Tech University

∨

2010
服饰品设计专业
Fashion Accessory Design

Branch Of Learning
学院开设

服装设计与工程
Fashion Design and Engineering

服装与服饰设计(服装艺术设计、服饰品设计)
Fashion and Accessory Design (Fashion Art Design,
Fashion Accessory Design)

产品设计(纺织品艺术设计)
Product Design (Textile Art Design)

表演(时装表演艺术、人物形象设计)
Performance (Fashion Performance Art, Fashion and Beauty Design)

中美合作项目
Sino-American Cooperation Project

服装设计与工程
Fashion Design and Engineering

服装与服饰设计(服装设计、设计与营销)
Fashion and Accessory Design (Fashion Design, Fashion Design and Marketing)

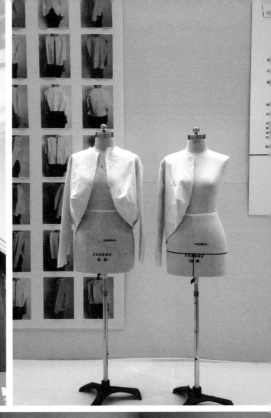

中国著名时装学府
Chinese famous fashion institution of higher learning

创新创业能力 高素质服装人才
Creative Ability and Entrepreneurship,
High-quality Talents of Fashion Industry

艺术设计与工程技术相结合
Combination of designing and engineering

创意设计与产品设计相结合
Creative design and product design

学校教育与社会实践相结合
School education and practical work

国内教学与国际合作教学相结合
China's teaching and international cooperative education

培养创新精神和实干精神，为服装企业孵化和储备设计人才
Training students of creative spirit and executive ability,
to develop design talents for fashion industry.

探索设计的可能性
Explore Design Possibilities

项目式教学
Project Teaching

梯队式教学团队
Teacher Echelon

零距离对接服装企业
Integrate Into Fashion Industry

引入式教学
Case-based Instruction

卓越人才培养计划
Excellent Talents Trainee Program

1

服 装 与 服 饰 设 计
❧ 服装艺术设计 ❧
Fashion and Accessory Design
—— Fashion Art Design ——

4

产 品 设 计
❧ 纺织品艺术设计 ❧
Product Design
—— Textile Art Design ——

2

中 美 合 作 项 目
服 装 与 服 饰 设 计
❧ 服装艺术设计——服装设计与营销 ❧
Sino-American Cooperation Project
Fashion and Accessory Design
—— Fashion Art Design ——
—— Fashion Design and Marketing ——

5

表 演
❧ 时装表演艺术 ❧
Performance
—— Fashion Performance Art ——

3

服 装 与 服 饰 设 计
❧ 服饰品设计 ❧
Fashion and Accessory Design
—— Fashion Accessory Design ——

6

表 演
❧ 人物形象设计 ❧
Performance
—— Fashion and Beauty Design ——

服装与服饰设计
❀ 服装艺术设计 ❀
Fashion and Accessory Design
—— Fashion Art Design ——

服装与服饰设计
服装艺术设计

Fashion and Accessory Design
— Fashion Art Design —

本专业方向依据服装行业及学科发展
人才需求方向，培养具有较强艺术素养、综合艺术
创新创造设计能力和设计研究分析能力，具有市场意识和眼光、
团队协作及沟通能力、实践操作能力和动手能力强的应用型、复合型服
装设计专业人才。本专业设有服装设计学科基础专业理论、专业知识、
专业技能技术课程，服装艺术设计修养课程，以及研究型服装项目设计
课程，模块系列服装设计课程。专业课程设计系统完善，使学生具有从
事服装艺术创新设计、女装系列产品设计、男装系列产品设计、针织服
装设计、服装品牌商品设计与企划，服装品牌形象设计策划、服装终端
陈列与推广、服装品牌营销及推广、流行时尚传播及推广、流行信息及
市场信息情报分析、流行市场的调研与分析、服装设计专业教育培训、
服装设计管理的能力。

The major direction follows the talents demand direction of Fashion industry and subject development, and cultivates compound practicability fashion design specialized talents possessing relatively high art accomplishment, comprehensive art innovation and creation design ability and design research analysis ability, possessing market consciousness and eyes, team coordination and communication ability, practical manipulation ability and relatively high hands-on ability. The major opens foundation major theory, major knowledge, major skill and technology courses of Fashion design subject, Fashion art design training courses, investigative Fashion project design courses, and module series Fashion design courses. The design of major courses is systematic and perfect, which enables students to have the ability of specializing in work such as Fashion art innovation design, women's wear series products design, men's wear series products design, knitted Fashion design, commodity design and planning of Fashion brands, design and planning of Fashion brand image, display and promotion of Fashion terminal, marketing and promotion of Fashion brand, fashion broadcasting and promotion, analysis of fashion information and market information, investigation-survey and analysis of fashion market, professional education training of Fashion design, and Fashion design management.

CLOTHING
TECHNOLOGY
LABORATORY
服装工艺实验室

CLOTHING
PLATE PUBLIC
SPACE
服装制板公共空间

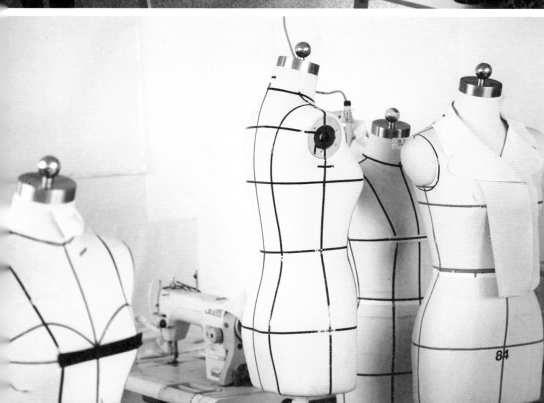

DRAPING
CLOTHING STUDIO
服装立体裁剪工作室

2

中美合作项目——服装与服饰设计
服装艺术设计——服装设计与营销
Sino-American Cooperation Project
Fashion and Accessory Design
Fashion Art Design
Fashion Design and Marketing

服装与服饰设计
中美合作项目
服装艺术设计
服装设计与营销

Fashion and Accessory Design
Sino-American
Cooperation Project
—— Fashion Art Design ——
Fashion Design and Marketing

本专业方向面向全球服装产业,培养具有国际化视野、具有敏锐的时尚嗅觉和市场感知力、具有较高的艺术修养和创作能力,掌握服装艺术设计领域的知识、理论和技能,能独立进行时装设计、服饰商品开发及相关的艺术设计及管理的高级服装设计人才。毕业生能从事时装设计、服装样板设计、服装企业商品计划、经营管理、营销策划、服饰品设计、广告设计及时装杂志编辑等工作。本专业是获国家教育部批准的中外合作办学项目,引进美国纽约州立大学时装技术学院(FIT)专业课程体系和教材,三分之一课程使用原版英文教材并由美方教师亲临本校授课。

The major direction faces global fashion industry and cultivates high-grade fashion design talentswho possess international viewpoint, possess sharp fashion smell and market sensibility, possess relatively high art accomplishment and creation ability, grasp knowledge, theory and skill of Fashion art design field, and can independently carry out fashion design, accessory commodity development, and relative art design and management. The graduates can work in positions such as fashion design, Fashion pattern design, Fashion industry commodity planning, operation and management, marketing planning, accessory design, advertising design and fashion magazine editor etc. The major is the sinoforeign cooperation education project which is approved by national education ministry. It imports major course system and textbook of New York State University Fashion Institute of Technology (FIT), and one third of the courses use original edition textbooks and are taught by American teachers who come to our school to teach in person.

FASHION DRAWING TEACHING
时装绘画教学

面料创新 工艺技法
廓形表达 时尚解读
服装艺术设计是一门艺术，也是一门手艺
是仰望星空的想象力
也是脚踏实地的执行力

3

服 装 与 服 饰 设 计

✤ 服饰品设计 ✤

Fashion and Accessory Design

— Fashion Accessory Design —

服装与服饰设计
服饰品设计
Fashion and Accessory Design
— Fashion Accessory Design —

本专业方向依据现代服饰产业市场细分化、专业化发展人才需求，培养具有较强艺术素养、审美能力、创新意识、综合艺术创新创造设计能力，掌握相关服饰设计的专业知识、专业技术技能，具有市场意识和眼光、团队协作及沟通能力，有较强的实践操作能力和动手能力，能从事服饰品牌策划与服饰产品设计企划、时尚鞋类产品创新设计、时尚包类产品创新设计、创新性饰品设计、服饰贸易、服饰品市场调研及服饰流行资讯分析、服饰品陈列与展示等综合素质高的复合型应用性服饰设计人才。

The major direction follows the segmentation and professionalization talents demand direction of modern accessory industry development, and cultivates students possessing relatively high art accomplishment, aesthetic ability, innovation consciousness, comprehensive art innovation and creation design ability, and possessing major knowledge, major skill and technology of related accessory design, with market consciousness and eyes, team coordination and communication ability, practical manipulation ability and relatively high hands-on ability. The major cultivates compound practicability fashion design specialized talents with high compound quality who specialize in work such as accessory brand planning, accessory commodity design and planning, innovative design of fashionable shoes products, innovative design of fashionable bags products, innovative accessory design, accessory trade, investigation-survey and analysis of accessory market, analysis of accessory fashion information, and accessory display and presentation etc.

服装设计是黄钟大吕的高妙和谐
服饰品设计如环佩叮当般活色生香

产 品 设 计
✤ 纺织品艺术设计 ✤
Product Design
— Textile Art Design —

产品设计
纺织品艺术设计
Product Design
— Textile Art Design —

本专业方向培养具有一定国际视野、现代设计理念、适应社会经济发展需求的纺织品艺术设计高级应用型人才，具有较强审美意识和市场意识、艺术设计表现能力，掌握纺织品艺术设计理论及设计专业技能技术，系统掌握图案设计理论、图案设计方法及设计表现，熟练掌握计算机图形运用技术，有较强创新创业精神、创意设计能力和实践操作能力，具有纺织品市场调查分析能力；专业知识面宽，基础扎实，能运用多学科的知识从事各类纺织品艺术创造性设计，其中包括服饰纹样设计、服饰织花提花设计、家纺织花提花和印花设计、纤维艺术设计、刺绣手绘艺术设计、纺织品面料开发设计、纺织品陈列展示、室内装饰配套设计、家纺产品造型与设计、纺织品设计管理。

The major direction cultivates high-level practicability talents of textile art design who possess certain international field of view, modern design idea, and adapted to social and economy development demand. It cultivates students possessing relatively high aesthetic consciousness and market consciousness, and expression ability of art design; grasping textile art design theory and professional design skill and technology; systematically grasping pattern design theory, pattern design method and design representation; proficiently grasping manipulation technology of computer graphics; having relatively high innovation and start-undertaking spirit, originality design ability, and practical operation ability; possessing investigation and analysis ability of textile market, with wide major knowledge and sturdy basis. The major cultivates students who can use multi-subject knowledge to specialize in all types of textile art creativity design, including costume pattern design, costume woven-pattern and jacquard-weaving design, woven-pattern, jacquard-weaving and print design of home textile, fiber art design, embroidery hand-drawing art design, textile and fabric development design, textile display and presentation, interior decoration combined design, home textile product modeling and design, and textile design management.

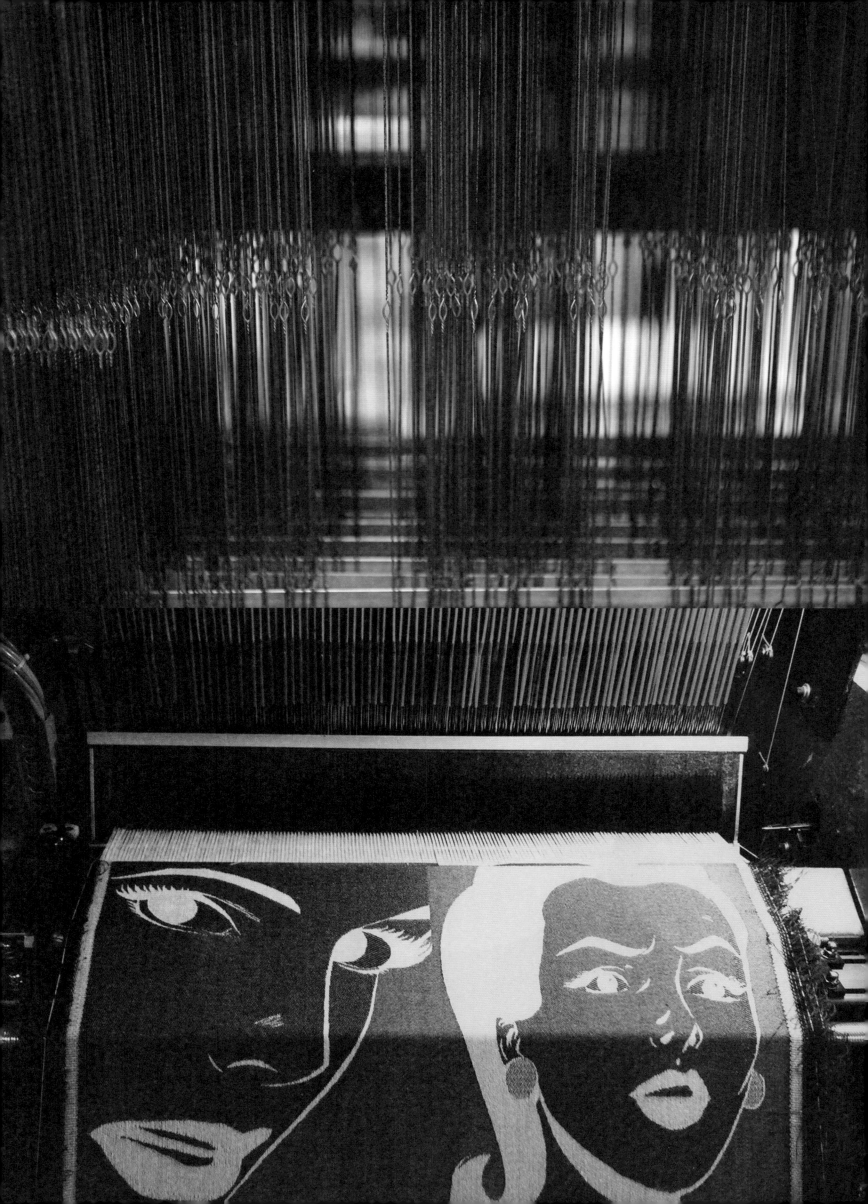

织物设计与工艺实验室

用染料绘成的画作
以纤维写就的诗篇
经纬间织进人间百态
纹样里描摹大千世界

表 演
时 装 表 演 艺 术
—— Performance ——
Fashion Performance Art

本专业方向依据时尚行业及时尚衍生产业人才需求特点，培养具备良好艺术气质及形象、优美形态，具有一定的艺术修养及艺术表演能力，掌握服装表演展示技巧及技能、了解时尚行业市场趋势动向，具有一定的时尚形象设计、服装形象搭配设计、服装货品陈列及组合设计、服装品牌终端营销推广等基本专业知识及能力，有较强的创新精神和组织策划能力、沟通能力、团队协作能力，能从事时装表演、时装表演策划及编导、时装表演教育与培训、时尚形象设计与策划、时尚经纪管理、时尚传播及推广、时尚活动展示经营策划、服装形象搭配设计、服装货品陈列及构成设计的高素质应用型综合性时尚人才。

The major direction is in accordance with the characterristics of talents demand for fashion industry and fashion derivative industry to cultivate students possessing good art temperament, figure, and graceful shape, possessing certain art accomplishment and art performance ability, grasping fashion performance and display skill and technique, and knowing trend and movement of fashion industry and market. It cultivates students possessing certain basic major knowledge and ability of fashion image design, fashion image matching design, display and combination design of Fashion products, and terminal marketing and promotion of Fashion brand, with relatively high innovation spirit, organization and planning ability, communication ability and team coordination ability. The major cultivates high-quality practicability and comprehensive fashion talents who can specialize in fashion performance, planning and director of fashion performance, education and training of fashion performance, design and planning of fashion image, fashion broker management, fashion broadcasting and popularization, operation and planning of fashion activities, fashion image matching design, display and constitution design of Fashion product.

6

表　演

✤ 人 物 形 象 设 计 ✤

Performance

—— Fashion and Beauty Design ——

表演
人物形象设计
—— Performance ——
Fashion and Beauty Design

本专业方向基于时尚产业变化发展趋势对时尚人物形象设计创新型应用型人才的需求，与韩国水原女子大学合作，培养具有较强的审美意识、艺术修养、艺术设计思维能力和创造能力、独立创新设计与实践能力、设计研究能力，沟通能力及团队协作能力；掌握人物形象设计理论、设计方法、设计技术技能，进行人物造型创新设计、时尚形象创新设计；掌握一定的服饰设计与搭配的专业理论与专业技能，了解时尚领域流行趋势动向，能从事时尚人物形象设计、化妆设计、发型设计，从事影视、舞台形象设计、形象设计及策划顾问、美容美发形象设计教学培训等相关职业的人才。

Cooperated with Korea Suwon Women's University, the major is based on the demand of innovative and practical fashion human figure design talents under the developing trend of fashion industry changes. It cultivates students possessing relatively high art accomplishment, aesthetic ability, thinking ability and creation ability of art design, independent innovative design and practice ability, design and research ability, and team coordination and communication ability. The major enables students to possess human figure design theory, design methods, design skill and technology, to carry out human figure innovative design and fashion image innovative design, and to possess certain professional theory and professional skill of accessory design and matching, and get to know the fashion trend and movement in fashion field. It enables students to have the ability of specializing in work such as human figure design, make-up design, hair-style design, image design of movies and TVS and stage, image design and planning consultant, and teaching and training of beauty-hair salon image design.

PART-2

一件服装的生长

从 种 子 到 结 果

A Process of Fashion Works
—— From Idea to Creation ——

创意是一个小小的种子

落进泥土里

看看它如何萌芽

有些喜阴　有些耐寒

有些需要更多的雨露　有些终究会夭折

就像我们的作品

有美好　有玩味　有挣扎　有成长

Creations are like little seeds
Fallen into the earth
Let's see how they sprout
Some of them are conditioned to shade, some can resist cold
Some need more rain, and some would die young
They are just like our composition.
They are happiness and fun, struggling and developing as well

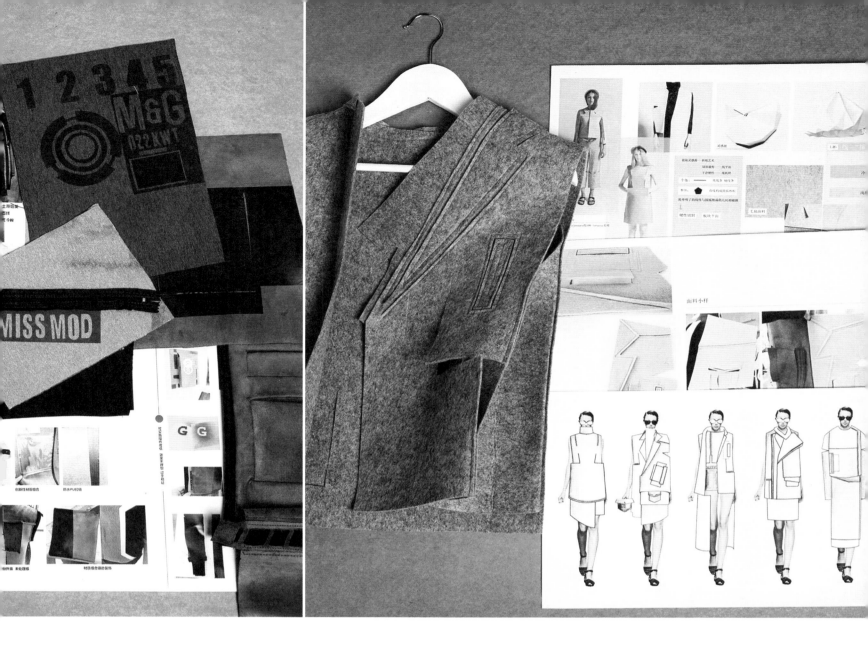

把创意画出来，时装是平面的；把面料缝制起来，时装是立体的
把服装挂起来，时装是静止的；把服装穿在身上，时装是流动的
构思创意 - 面料改造 - 小样制作 - 结构探索 - 廓形表达
服装，是一件装置艺术；时装，是一种行为艺术

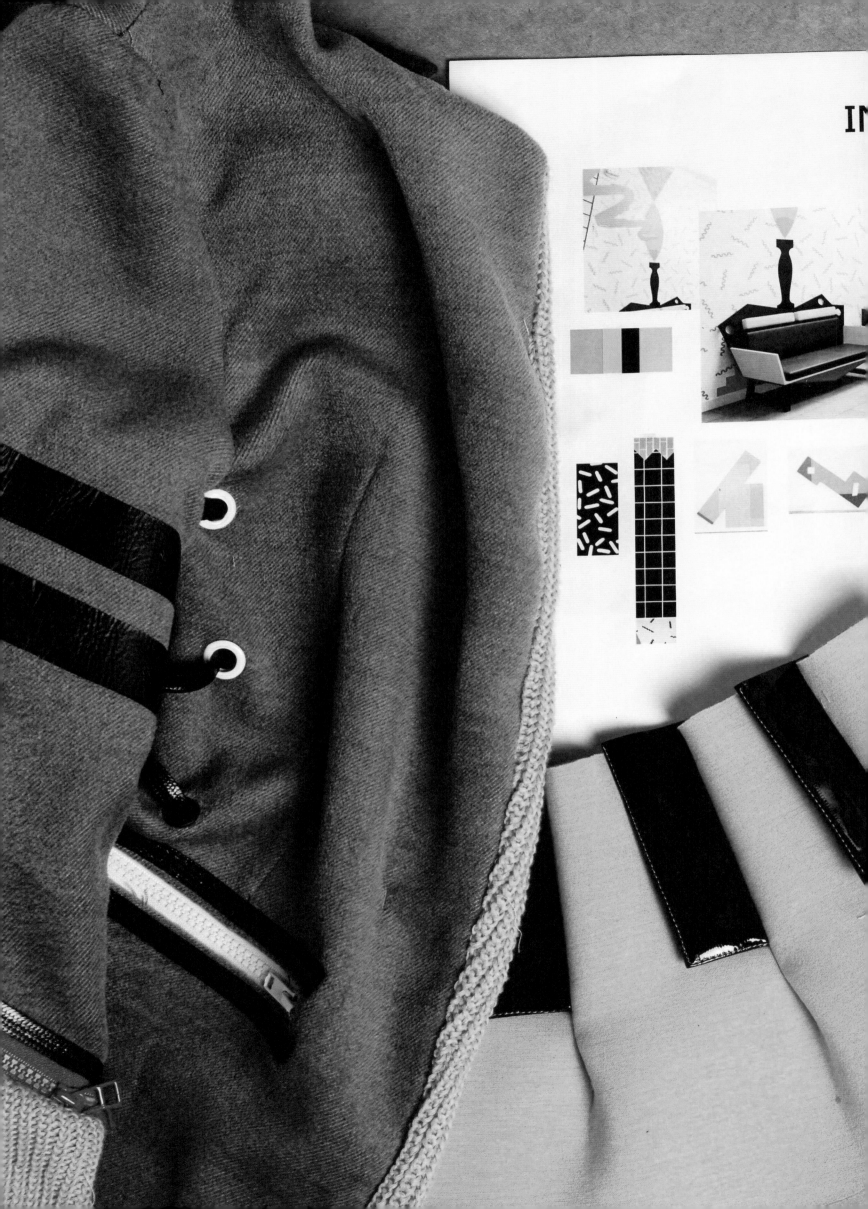

OCENT·天真

关键词-调侃 生动

趣味化运动化的正装

设计师: 李玲莉　Designer : Li Lingli　|　指导老师: 陈敬玉　Director : Chen Jingyu

设计师：张雁　Designer：Zhang Yan　|　指导老师：陈敬玉　Director：Chen Jingyu

设计师：林晨晨　　Designer：Lin Chenchen

指导老师：陈敬玉　　Director：Chen Jingyu

设计师: 陆成文　　Designer : Lu Chengwen
指导老师: 陈玲芳　Director : Chen Lingfang

设计师: 项鑫　　Designer : Xiang Xin
指导老师: 陈玲芳　Director : Chen Lingfang

设计师：卜世鹏　　Designer：Bu Shipeng
指导老师：陈玲芳　　Director：Chen Lingfang

设计师：汤柠霜　　Designer：Tang Ningshuang
指导老师：陈玲芳　　Director：Chen Lingfang

设计师: 张雁
Designer : Zhang Yan

指导老师: 陈敬玉
Director : Chen Jingyu

设计师: 潘文洁　　Designer : Pan Wenjie
指导老师: 须秋洁　　Director : Xu Qiujie

设计师: 李思缌　　Designer : Li Sisi
指导老师: 陈玲芳　　Director : Chen Lingfang

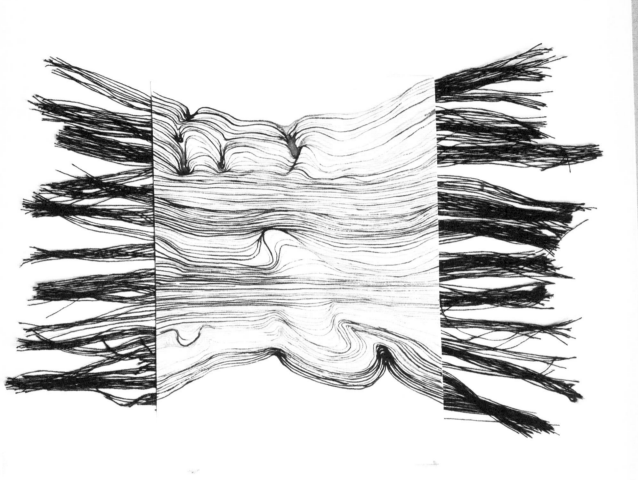

T eight

PART-3

·一 个 创 意 的 实 现·

我 和 我 的 作 品

Graduation Work
—— Collection ——

在这里

有我对生活的最初理解

对人群的最初想象

是奔走市场疲惫的双脚

是熬夜赶工通红的睡眼

有惊喜 有沮丧

欢乐和悲伤

青春 就是要将时间浪费在美好的事物上

所以 浪费的时间和美好的事物组成了

我的大学

Here is the my first comprehending of life

My first imagination of the customer

Here are exhausted feet for market research

Sleepless eyes that work and stay up late

Surprise, depression, glorious and sadness

Youth time is to be wasted in the wondrous things

Therefore, the wasted time and the wondrous things composed

My graduated works

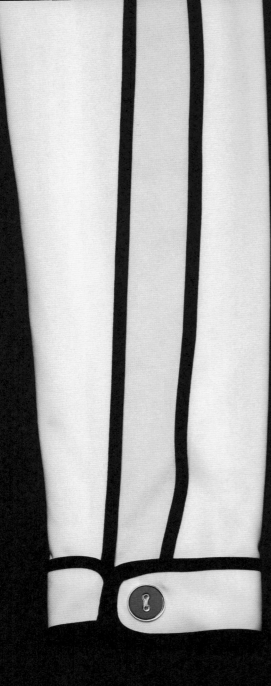

设计师：贾晓敏　Designer：Jia Xiaomin

指导老师：陶宁　Director：Tao Ning

设计师: 戴梦瑜   Designer : Dai Mengyu | 指导老师: 贺华洲   Director : He Huazhou | 模特: 谢腾   Model : Xie Teng

设计师: 蒋优洋   Designer: Jiang Youyang
指导老师: 钮敏   Director: Niu Min
模　特: 谢腾   Model: Xie Teng

设计师: 张蔻华　Designer : Zhang Kouhua　|　指导老师: 陶宁　Director : Tao Ning　|　模特: 谢腾　Model : Xie Teng

设计师: 程宁杰　Designer : Cheng Ningjie

指导老师: 陶宁　Director : Tao Ning

模 特: 谢腾　Model : Xie Teng

设计师：李冉   Designer：Li Ran | 指导老师：陶宁   Director：Tao Ning | 模特：谢腾   Model：Xie Teng

设计师：刘淑　　Designer：Liu Shu
指导老师：夏帆　　Director：Xia Fan
模　特：赵庆贺　　Model：Zhao Qinghe

设计师：刘淑　Designer：Liu Shu　|　指导老师：夏帆　Director：Xia Fan　|　模特：赵庆贺　Model：Zhao Qinghe

设计师：姜诗丽   Designer：Jiang Shili
        楼沙漠霜   Lousha Moshuang

指导老师：陆珂琦   Director：Lu Keqi

模 特：赵庆贺   Model：Zhao Qinghe

设计师：姜诗丽   Designer：Jiang Shili
        楼沙漠霜   Lousha Moshuang

指导老师：陆珂琦   Director：Lu Keqi

模 特：赵庆贺   Model：Zhao Qinghe

设计师: 姜诗丽 / 楼沙漠霜　Designer: Jiang Shili / Lousha Moshuang　|　指导老师: 陆珂琦　Director: Lu Keqi　|　模　特: 赵庆贺　Model: Zhao Qinghe

设计师：袁舒宜　　Designer：Yuan Shuyi
指导老师：陆珂琦　　Director：Lu Keqi
模　特：赵庆贺　　Model：Zhao Qinghe

设计师: 袁舒宜　Designer : Yuan Shuyi　｜　指导老师: 陆珂琦　Director : Lu Keqi　｜　模特: 赵庆贺　Model : Zhao Qinghe

设计师: 王予琴
Designer : Wang Yuqin

指导老师: 姚琛
Director : Yao Chen

设计师: 罗铮   Designer: Luo Zheng | 指导老师: 胡琼   Director: Hu Qiong

设计师：徐凌燕
Designer：Xu Lingyan

指导老师：朱寒宇 / 顾小燕
Director：Zhu Hanyu / Gu Xiaoyan

设计师: 郭佳祥
Designer: Guo Jiaxiang

指导老师: 陈敬玉
Director: Chen jingyu

设计师: 张音培
Designer : Zhang Yinpei

指导老师: 冯荟
Director : Feng Hui

设计师: 姚魏煊
Designer : Yao Weixuan

指导老师: 凌雯
Director : Ling Wen

设计师：张炜　　Designer : Zhang Wei
指导老师：严昉　　Director : Yan Fang
模　特：祝淑敏　　Model : Zhu Shumin

设计师：王维　Designer：Wang Wei　|　指导老师：严昉　Director：Yan Fang　|　模特：梁佳一　Model：Liang Jiayi

设计师: 刘小丽
Designer : Liu Xiaoli

指导老师: 陈玲芳
Director : Chen Lingfang

模 特: 徐瑾
Model : Xu Jin

设计师: 张沥尤　Designer: Zhang Liyou　|　指导老师: 陈敬玉　Director: Chen Jingyu　|　模特: 徐瑾　Model: Xu Jin

设计师：夏雨田
Designer : Xia Yutian

指导老师：须秋洁
Director : Xu Qiujie

模 特：谢锦阳
Model : Xie Jinyang

设计师：高莹莹　Designer：Gao Yingying　|　指导老师：须秋洁　Director：Xu Qiujie　|　模特：谢锦阳　Model：Xie Jinyang

设计师：邱耶轶轶　Designer：Qiuye Yiyi
指导老师：陈敬玉　Director：Chen Jingyu
模　特：祝淑敏　Model：Zhu Shumin

设计师: 王元可　Designer : Wang Yuanke　|　指导老师: 须秋洁　Director : Xu Qiujie　|　模特: 谢锦阳　Model : Xie Jinyang

STAFF LIST
浙江理工大学
服装学院
工作人员名单

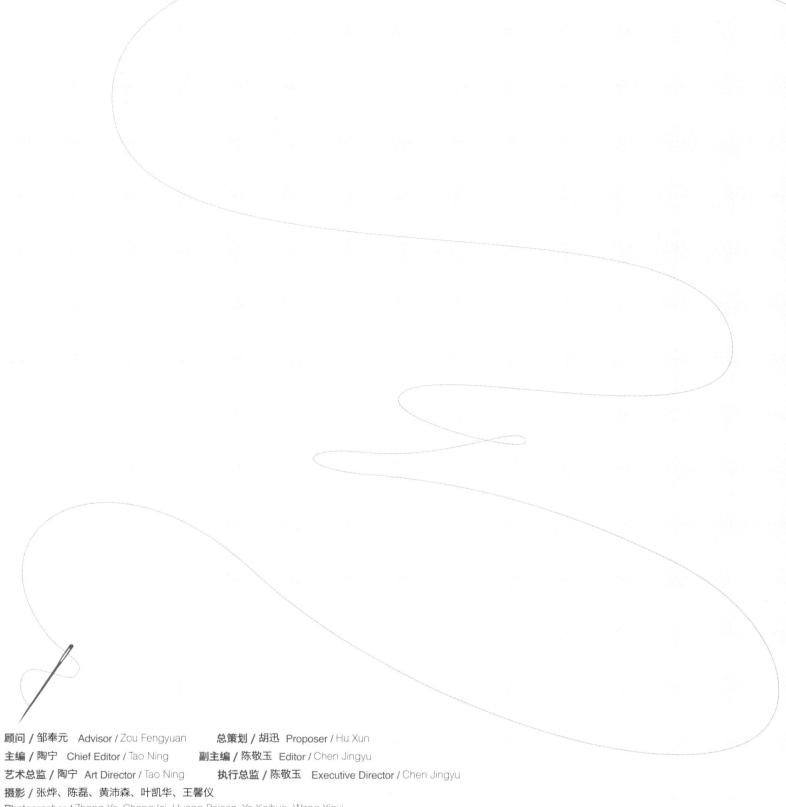

顾问 / 邹奉元　Advisor / Zou Fengyuan　　　总策划 / 胡迅　Proposer / Hu Xun

主编 / 陶宁　Chief Editor / Tao Ning　　　副主编 / 陈敬玉　Editor / Chen Jingyu

艺术总监 / 陶宁　Art Director / Tao Ning　　　执行总监 / 陈敬玉　Executive Director / Chen Jingyu

摄影 / 张烨、陈磊、黄沛森、叶凯华、王馨仪
Photographer / Zhang Ye, Cheng lei, Huang Peisen, Ye Kaihua, Wang Xinyi

造型与化妆 / 张天一、麻湘萍、元晓敏、尹雅兰、李东
Image Designer / Zhang Tianyi , Ma Xiangping, Yuan Xiaomin, YOON A RAM, Li Dong

服装统筹 / 胡迅、陶宁、陈敬玉　Garment Coordinator / Hu Xun, Tao Ning, Chen Jingyu

模特统筹 / 张瑾、郭亚楠　Model Coordinator/ Zhang Jin, Guo Yanan

平面设计 / 严磊、陈雷　Graphi Design / Neo, Chan Ray

文字及翻译 / 吕昉　Paperwork and translator/ Lv Fang

毕业设计总策划 / 赵卫国、夏帆、严昉、胡蕾、朱寒宇
Graduation Design Directors/ Zhao Weiguo, Xia Fan, Yan Fang, Hu Lei, Zhu Hanyu

指导老师 / 陈敬玉、陈玲芳、冯荟、顾小燕、胡琼、贺华洲、陆珂琦、李琴、凌雯、钮敏、钱丽霞、
陶宁、须秋洁、夏帆、严昉、姚琛、赵卫国、朱寒宇（按姓氏字母排名先后）
Design Advisors/ Chen Jingyu, Chen Lingfang, Feng Hui, Gu Xiaoyan, Hu Qiong, He Huazhou, Lu Keqi, Li Qin, Ling Wen, Niu Min, Qian Lixia, Tao Ning, Xu Qiujie, Xia Fan, Yan Fang, Yao Chen, Zhao Weiguo, Zhu Hanyu

特别鸣谢 / 黑耳视觉 / 龙腾精英国际模特经纪（北京）有限公司 / DBC 榴莲品牌策划有限公司
Special Acknowledgement/
Herevision
Longteng Supermodel Agency International (Beijing) CO.,Ltd
Durian Brand Consultants Co.Ltd